Mohammad Rezaul Karim

Analysing the Role of Triangulation in Research

GRIN Verlag

Bibliografische Information der Deutschen Nationalbibliothek:

Die Deutsche Bibliothek verzeichnet diese Publikation in der Deutschen National-
bibliografie; detaillierte bibliografische Daten sind im Internet über http://dnb.d-
nb.de/ abrufbar.

Dieses Werk sowie alle darin enthaltenen einzelnen Beiträge und Abbildungen
sind urheberrechtlich geschützt. Jede Verwertung, die nicht ausdrücklich vom
Urheberrechtsschutz zugelassen ist, bedarf der vorherigen Zustimmung des Verla-
ges. Das gilt insbesondere für Vervielfältigungen, Bearbeitungen, Übersetzungen,
Mikroverfilmungen, Auswertungen durch Datenbanken und für die Einspeicherung
und Verarbeitung in elektronische Systeme. Alle Rechte, auch die des auszugsweisen
Nachdrucks, der fotomechanischen Wiedergabe (einschließlich Mikrokopie) sowie
der Auswertung durch Datenbanken oder ähnliche Einrichtungen, vorbehalten.

Imprint:

Copyright © 2007 GRIN Verlag GmbH
Druck und Bindung: Books on Demand GmbH, Norderstedt Germany
ISBN: 978-3-656-36834-2

This book at GRIN:

http://www.grin.com/en/e-book/207958/analysing-the-role-of-triangulation-in-
research

GRIN - Your knowledge has value

Der GRIN Verlag publiziert seit 1998 wissenschaftliche Arbeiten von Studenten, Hochschullehrern und anderen Akademikern als eBook und gedrucktes Buch. Die Verlagswebsite www.grin.com ist die ideale Plattform zur Veröffentlichung von Hausarbeiten, Abschlussarbeiten, wissenschaftlichen Aufsätzen, Dissertationen und Fachbüchern.

Visit us on the internet:

http://www.grin.com/

http://www.facebook.com/grincom

http://www.twitter.com/grin_com

Analysing the Role of Triangulation in Research

M. Rezaul Karim[*]

Abstract

Research is always conducted to find out solution(s) of a problem. There are so many factors involved in the research process from data collection to result analysis. These factors sometimes distort the result, sometimes influence the findings. But a valid and reliable result is always wanted and widely acceptable. To make the research result bias free, valid and generalised triangulation plays an important role in this area by increasing the rate of certainty and bringing neutrality. It is process of using more than one method, theory, researcher and data collection method & technique to make the research findings more valid, reliable and generalisable.

1.0 Introduction:

Triangulation is the means of reducing bias in research and increases the rate of certainty of the research findings. It goes without saying that triangulation is very important in research methodologies. The beneficiary groups or incumbent of the research topic want findings bias free, reliable even tested in the long run. Researchers do for the people following some techniques, process from data collection to interpretation, analysing those collected data. The whole process involves many things to be done.   In every step, for the component neutrality is necessary. And this is the issue of triangulation. It follows the means how to reduce partiality and raise neutrality and certainty. Tomlinson (2006) remarks it serves as a check on the validity and reliability. In the research triangulation indicates the real position of research. It involves locating a true position by referring to two or more other coordinators Densccombe (2003:133).

Reviewing literatures there are four approaches of triangulation are found. In every stage it is necessary to avoid bias and should unitise authentic data and information that will certainly lead a research to acceptable conclusion to everybody.

This essay aims to discuss the importance of all forms of triangulation approaches in research methodologies and examine the importance thereof and how they play vital role for reliable and valid research findings.

2.0 Triangulation

It is a process of overcoming bias and developing certainty in the research methodology checking data validity, reliability, theoretical issues, interviewers' biasness and methodological problems. In terms of significance Jick (1979) and Thietart et al (2001:83) discusses 'triangulation allows the researcher to benefit from the advantages of the two approaches, counterbalancing the defects of one approach with qualities of the other.' Denzin (1978) defines triangulation as 'the combination of methodologies in the study of the same phenomenon'. 'Introducing triangulation into research design is one means whereby the evidence collected from one source is corroborated by evidence collected from another source , with the discrepancies emerging between the two sets of data altering researchers to potential analytical errors. Thus, triangulation can enhance our belief that results are valid and not a methodological artefact.' Bouchard (1976) and Adam and Healy (2000: 58). Ghauri, *et al* (1995:94) defines triangulation as the combination of methodologies in the

---

[*] PhD Student at Graduate School of Public Administration (GSPA) , National Institute of Development Administration (NIDA), Bangkok, Thailand and Permanent Faculty Member of Bangladesh Public Administration Training Centre (BPATC), Savar, Dhaka-1343, Bangladesh.

study of a same phenomenon. So it is seen that every author defines the triangulations as the combination of different approaches and methods in the same phenomenon that helps the researcher to overcome bias and uncertainty in the research findings in order to be widely acceptable and useful for the future research.

Triangulation is system of using more than one method where all methods interrelated to provide dependable research result. Decommbe (2000) shows the true location of triangulation discussing the interrelation between methods

According to Esterby-Smith, et al (2002), Denzin (1970) in Thomas (2004:23) four types of triangulation are identified.

    2.1 Data Triangulation: Here data is collected from different sources in the study of a phenomenon.

    2.2 Theoretical Triangulation: Theoretical triangulation is approach where it is taken from a discipline to describe the phenomenon of other discipline. Esterby-Smith , et al (2002:146), Thorpe and Lowe say that 'Theoretical triangulation involves borrowing models from one discipline and using them to explain situations in another discipline.'

    2.3 Triangulation by Investigator: Here different researchers collect data individually and independently for the same phenomenon. They compare the findings on the basis of collected data.

    2.4 Methodological Triangulation: It is the way of using of the both qualitative and quantitative methodologies in research in order to get reliable findings. This comprises of within the method triangulation and between the method triangulation.

3.0 Importance of triangulation in Research

Triangulation has a significant role in research methodology to prove the research as an important, viable and widely accepted. It brings validity, increased methodological reliability and the rate of certainty in the research findings. It happens affecting by the all forms of triangulation used in the research.

Forms of triangulation that affect research findings leading to acceptable results: Theme has been taken from Kane and Brun (2001)

In the triangulation in research design researchers use the evidences collecting from one to another source and minimize data errors. Thus, triangulation can enhance our belief that results are valid and not a methodological artefact. Bouchard, 1976 cited in Adam and Healy (2000:58). Denzin (1970) cited in Hussey and Hussey (1997:74) argues that the use of different methods by a number of researchers studying the same phenomenon should, if their conclusions are the same, lead to greater validity and reliability than a single methodological approach. It offers a balance between logic and stories. (http://www.ucalgary.ca/ ~dmjacobs/phd/methods)

It helps the researchers to choose relevant data collection methods, minimises uncertainty reducing bias, minimising personal affects on the research findings. 'However, it is important that the research question is clearly focused and not confused by the methodology, adopted and that the methods are chosen in accordance with their relevance to the topic.' (Ticehurst and Veal ,2000:51). As the triangulation 'can produce a more complete, holistic and contextual portrait of the object under the study' (Ghauri, et al ,1995:94) it provides advantage to the research. Even it can be useful technique in complex phenomena. (Cohen and Manion, 1989:277)

Triangulation is employed for a number of reasons. Sarantakos (1998) shows the reasons and summarises its importance in the research. It allows the researcher:
- to obtain a variety of information on the same issue;
- to use the strength of each method to overcome the deficiencies of the other;
- to achieve a higher degree of validity and reliability; and
- to overcome the deficiencies of single-method studies.

Reviewing literature it is seen that there some importance of triangulations that can be categorised into six points.
1. Triangular techniques are suitable when a more holistic view is sought in research. Most research of this kind looks at an achievement or skill outcome rather than the development of attitudes.
2. Triangulation has special relevance where a complex phenomenon requires elucidation. Because of the contrasting philosophies, objectives and practices in the two classes, single method provides limited value, but the adaptation of multimethod approach would give very different features. Can get more realistic view.
3. It is appropriate when different methods of learning are to be evaluated. Skills criteria can be found here.
4. It is suitable for controversial aspect of research where needed to be evaluated more fully. Here these could measure and investigate factors such achievement, teaching methods, practical skills, cultural interests, social skills, interpersonal skills, community spirit and so on. Validity could be then increased.
5. It is useful when an established approach yields a limited and frequently distorted picture.
6. It can be useful technique where a researcher is engaged in case study particularly examining a complex phenomenon. (Cohen and Manion 1989:275-7)

3.1 *Data Triangulation*: It is a process of using more than one data collection technique to make the research findings more reliable and variable. For the research methodology data is the main factor to influence the research result. In terms of data collection there is possibility to collect data from less important sources, even less important data. If the researchers are having no experience their data collection may be affected by the ignorance of interviewees who may give wrong data, may be given for not understanding the question or for personal problem. Anyway these data will certainly affect the research findings. Primary data is time consuming and sometimes interviewees do not give due attention. For this reason data may be biased or incorrect that will definitely affect research findings. So data triangulation can play vital role in research methodology to get reliable result minimising data errors. Getting the primary data from the source, from an authentic source data triangulation can reduce bias. If the data collected from genuine, authentic sources research findings will be more acceptable no doubt. It would be more reliable and valid. So data triangulation proves that it is very much important in research methodology. If data triangulation is maintained, validity and reliability will be higher in the research that will lead to acceptability. To do this data triangulation examines the influence past and present time, different type of people and their level of difference and even the situation used from the different side. Everything depends on the perception and use of researchers how and when they use Blumberg *et al* (2005). Kane and Brun (2001:110) also remark that because of the use of different time and situation research may provide different results. To get more significant result investigators should emphasize on these everything in terms of data collection.

3.1.1 *Reliability:* 'Reliability relates to the methods of data collection and the concern that they should be consistent and not distort the findings. Generally it entails an evaluation of methods and techniques and to collect the data' Denscombe (2000:100). So data triangulation increases the rate of reliability in research and provides generalised result.

3.1.2 *Validity:* Validity concerns the accuracy of the questions asked the data collected and the expectations offered. Generally it relates to data and the analysis used in the research (Denscombe, 2000:100). As more than one data collection technique is used in this process it increases the rate of validity in research findings.

3.1.3 *When it occurs:*
Commentators remark (1998 Philip, 1971 and Sarantakos,) that data triangulation occurs when the researchers:

- Examine the influence of different times. Past and present on whatever they are studying.
- Examine the influence of space , that is compare the data with from other place (other organisations, others fields, and so on)
- Examine the person at different levels; the individual level, the group level, the collective level (entire group) level.
- Examine the situation from different angles; a field from the point of view of the head, the incumbents, the working staffs, the observers.

3.2 *Investigator Triangulation:* This is the system of collection data and analysing by more than one investigator that leads to same result in the research that proves that validity and reliability and subsequently acceptable to the general people to be used widely and furthermore. In the investigator triangulation 'different researchers independently collect data on the same phenomenon and compare the result' Hussey and Hussey (1997:74). If they do so, valid and reliable research findings can be achieved. But if they are bias in collecting data or do not pay full attention to the research or do not be punctual and sincere there will be a negative result on the research findings. Denzin (1978) and Hakim (1987) recommended multiple researchers as the only way of getting round the potential bias that comes from relying on a single researcher. The is the strength of research programmes.

It is necessary to reduce personal bias to dependable result in research. For this reason Researchers 'involve the researcher's dealing with management, respondents, and their own professional integrity' Fair Jr. *et al* (2003:104) and Kumar (1996). Researcher must be ethical in this connection to provide real issues everywhere that will produce valid findings. The ethical researcher always follows the analytical rules and conditions for results to be valid. The ethical researcher reports findings in many ways that minimize the drawing of false conclusion. The ethical researcher also uses chart, graphs and tables to show the data objectively, despite the sponsor's preferred outcomes Cooper and Schindler (2001:120). So it is seen that if the researchers become unethical, research findings will not be valid, reliable and acceptable.

Researchers have a significant role in conducting research to get dependable and applicable research findings. So investigator triangulation is very useful in research methodology for the certainty of research outcomes. Investigator can play a role in research methodologies keeping in mind about some validity threat that may distort the result. For the important role of the investigators there will be a minimum chance of biasness to be reflected on the research findings.

    3.2.1   **Reliability:** In the investigator triangulation different researchers works for the same phenomenon to get same research that brings reliable findings in the research.

    3.2.2   **Validity:** When many researchers collect information for the same phenomenon and get the same research, researcher's triangulation brings valid result in research methodologies.

    3.2.3   **Acceptability:** Investigator triangulation is the method to get the same result in a method explaining in different way by the different researchers. So in this case research is always acceptable by all as this can be used for future research.

3.3 *Theoretical Triangulation*: It is such a form of triangulation where one theory is taken from a discipline and is used to explain situation in another discipline. This proves that one theory is useful for the two or more condition and in various disciplines. In this perspective to get reliable and valid research result it is better to use induction and deduction approaches. Those will bring the certainty rate in the research findings. Kane and Brun (2001:110) remark that 'Theoretical triangulation is when you use different, or competing, theories to try to explain what is happening'

    3.3.1   **Reliability:** If the theory is tested in one discipline taken from another discipline and the positive result is gotten, the research findings will be more reliable.

    3.3.2   **Validity:** Theoretical triangulation develops the theoretical validity in the research methodology. And makes the result more acceptable.

    3.3.3   **Generalisability:** As the validity and reliability is maintained in theoretical triangulation, the research findings get the general acceptability. Researchers can use the theory for further use to develop theories.

3.4 *Methodological Triangulation*: Researchers use methodologies to find out means and ways in research design. They choose methodology on the basis of research topic what they will try to find out. Methodology is the main factor to conduct the research for finding out result. If the topic is fully based ton laboratory it is better to choose experimental method where quantitative data would be more useful and here internal validity will be high. And for the ethnographical methodology qualitative data is more useful. In this case external validity will be high. So thinking the research topic and aims of the research topic methodology should be identified and do the research on the basis on that. In this case methodological triangulation will be important to provide more reliable and valid research findings. Hussey and Hussey remark (2000:16) 'If you have a quantitative methodology you will be attempting to measure variables or count occurrences of a phenomenon. On the other hand, if you have qualitative methodology, you will emphasise meanings and experiences related to the phenomenon.'

Triangulation involves locating a true position by referring to two or more other coordinators.

Methodological Triangulation. Pp133

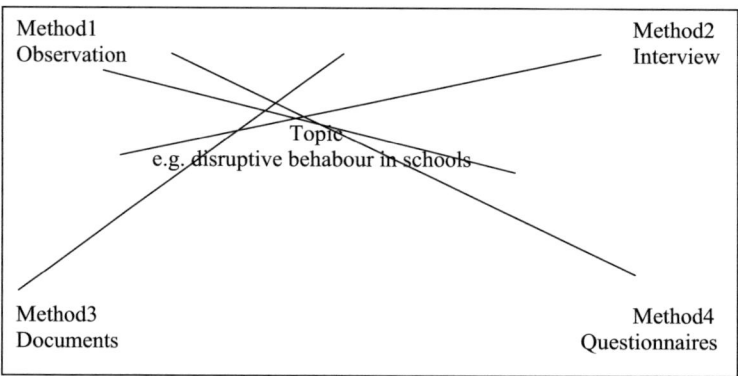

Source: The Good Research Guide written by Densccombe, M. (2003)

For minimising methodological biasness two types of methods are used for getting dependable result.

(a) *Single method*: Both qualitative and quantitative methods are used in research. But here it is seen that what will be more effective use of methods. Sometimes only qualitative or quantitative methods may be more viable to get acceptable result. On the basis of the data collection and research topic any method is chosen.

(b) *Multi-method*: Researchers always try to get the more reliable and valid research findings so that these can be widely accepted and useful for the further action or use. If they think they can get dependable and valid research findings using the combination of these two methods, multi-method is used then that helps to get reliable result from the research. Multi method can provide reliable findings than single method. Riley *at el* discussed the main advantages of multi-method approach (2000:39) 'The use of more than one research method to examine a particular phenomenon may improve understanding of that phenomenon and each technique may reveal facets of the phenomenon that would not be yielded by the use alternative methods.' In terms of importance of multi-method, Campbell and Fiske (1959) and Ghauri *et al* (1995:94) argues that 'to ensure validation one should use more than one method.'

3.4.1 Reliability and Validity: Using multi method in the research can provide more reliable and valid findings as both type of methods are

6

used here. When more than one method is used, the result is more tested and verified. So in this case this type of triangulation is very much useful for the acceptable research result.

3.4.2   Gereralisability: Triangulation is sometimes called as the multiple method as the multi method is used widely for minimizing the methodological biasness for a greater validity and reliability.

4.0 Critical Assessment of Triangulation and Limitation:

In triangulation if the threats to validity, historical testing, instrumentation, mortality, maturation, ambiguity about causal direction are not carefully handled (Saunder *et al, 2003*) these may distort research findings. Every method has some disadvantages that might distort the research to get bias free findings. 'There are some problems with triangulation. For example, sometimes it can be difficult to judge if the result from different methods are consistent or not. A second problem arises when the different methods come up with contradictory results. Sometime researchers prefer or emphasize one method over another: for example, quantitative versus qualitative approaches. However, all research methods have advantages and disadvantages when it comes to different research problems. (Ghauri, et al (1995:94). Philip (1971) remarks that human beings are not governed by scientific laws, so they are free to choose various course of action. And his behaviour will be affected by the environment, nature etc. Human behaviour is complex. Human beings cannot possibly study other human beings without bias. Because they bring values that may distort the data.   In terms of scientific methods it is the best intentions as to minimize the bias. Sarantakos (1998:169) also says that 'although the use of triangulation is generally thought to produce to more valid and reliable results than the use of single methods, there are some researchers who disagree. They argue that generalisations of this kind are unfounded and point to the fact that expanding the spectrum of methods employed to collect the does not necessarily guarantee better results.'

Triangulation   sometime brings various meaning in the research that makes a problem to utilise effectively in the research. Bryman remarks that triangulation has come to assume a variety of meanings although the association with the combined use of two or more research methods   within   a   strategy   .   (http://www.referenceworld.com/sage/socialscience/triangulation.pdf)

Wilson (1985) cited in Lecture delivered by Tomlinson (2006) discusses the purpose and procedure of triangulation that are thought as corroboration, elaboration and initiation.

Corroboration: One of the principal aims of triangulation in the social sciences seems to be corroborated one set of findings with another; the hope is that two or more sets of findings will 'converge' on a single proposition. This view holds much weight in literature on triangulation (http://www.voicewisdom.co.uk/htm/triangulation1.htm).

Elaboration: Triangulation provides a whole picture of the research that can be helpful for the researchers for the bias free findings. This is one of the most important aims of triangulation. It may occur through varieties of methods expanding upon our understanding of that which we research (Tomlinson, 2006)

Initiation: When the various forms of triangulation are utilised, sometimes potentials outcomes may come to do further research for our better understanding of the research. In the stage of utilizing triangulation in research methodologies new dimension may come that lead to take steps to adopt research again in the topic.

Measurement in triangulation is also a problem for which research result may vary. Different kinds of measurement can show the different pictures of result that depend on the situation. So proper utilisation of measurement in the triangulation can supply reliable and acceptable result.

During the use of methodological triangulation researchers should be more serious so that any kind of contradiction may not arise. Bowen (1996) remarks while methodological triangulation can enhance, illustrate, and clarify research findings, the researcher should keep in mind that use of multiple methods can also lead to the discovery of paradox and contradiction (http//www.bowensevaluation.com/ papers/Sin OfOmission/main.htm). There is an inevitable relationship between data collection methods employed and the result is obtained. The result will be off course affected by the method used. But it is the problem to ascertain the nature of that effect (Saunder *et al,* 2003:99-100)

5.0 Conclusion

In spite of some limitations various forms of triangulation are very much important in research methodology to make the research findings more reliable, valid and widely acceptable. As ' these involves analysing in more than one way , using more than one sampling strategy, using different interviewers, observers and analysts in the one study, and using more than one methodology to gather data.' Ticehurst and Veal (2000:50). Reliable data from reliable sources leads research to interpret main issues and helps to analyse critically to get real thing. Same way investigators play a vital role in research methodology to find out valid and reliable research results. If they are bias free, collect data accurately, become impartial research findings will definitely reliable and valid. Methodology is sometimes taken for a discipline, can be befitted for another discipline to explain. In this case researchers can easily utilise another methodology to get maximum output. In this case theoretical triangulation can assist a lot. In triangulation to improve confidence in the validity of data and findings the use of multiple approaches has increasingly been justified. Thomas (2004:23).

Here it can be concluded stating that triangulation or the use of a multi-method approach on the same study object can be useful even if we do not get the same results. It can lead us to a better understanding or to new questions that can be answered by later research Ghauri *et al* (1995:94).

Acknowledgement:

The author is very much grateful to Dr. Jenifer Tomlinson, Lecturer in HRM (Human Resource Management) of the University of Leeds, UK for her valuable comments on the draft essay.

References:
1. Adam, F. and Healy, M. (2000), *A Practical Guide to Postgraduate Research in the Business Area,* Ireland: Blackhall Publishing.
2. Bowen, K. A. (1996) *The Sin of Omission - Punishable by Death to Internal Validity: An Argument for Integration of Qualitative and Quantitative Research Methods to Strengthen Internal Validity (A PhD proposal)*, Cornell University. URL: (http//www.bowensevaluation.com/ papers/Sin OfOmission/main.htm)
3. Bouchard, T. J. (1976) 'Field Research Methods: Interviewing, Questionnaires, Participant Observation, Systematic Observation, Unobstrusive Measures', in M.D. Dunette (ed), *Handbook of Industrial and Organizational Psychology,* Chicago:Rand McNally.
4. Blumberg, B., Cooper, D. R. and Schindler, P. S. (2005) *Business Research Methods,* London: The McGraw-Hill Companies.
5. Campbell, D. T. and Fiske, D.W. (1959) Convergent and Discriminant Validation by the Multitrait-Multimethod Matrix, *Psychological Bulletin,* Vol. 56, p81-5.
6. Cohen, L and Manion, L. (1989), *Research Methods in Education (3rd Edn),* London: Routledge.
7. Cooper, D. R. and Schindler P.C. (2001) *Business Research Methods,* Singapore: McGraw-Hill International Edition.
8. Denscombe, M. (2002), *Ground Rules of Good Research: A 10 Point study of Social Research,* Buckingham & Philadelphia, pa: Open University Press.
9. Densccombe, M. (2003) *The Good Research Guide (2nd edn): for small-scale social research projects,* Maidenhead-Philadelphia: Open University Press.
10. Denzin, N. K. (1970) *The Research Act: A Theoretical Introduction to Sociological Methods* , Chicago: Aldine.
11. Denzin, N.K. (1978) *Sociological Methods: A Sourcebook,* New York: McGraw-Hill.
12. Easterby-Smith, M., Thorpe, R. and Lowe, A. (2002) *Management Research: An introduction,* London: Sage.
13. Fair Jr., J. F., Babin, B., Money, A. H. , Samouel (2003) *Essentials of Business Research Methods* , USA: Wiley.
14. Ghauri, P. N., Gronhaug, K. and Kristianslund, I (1995) *Research Methods in Business Studies: A practical Guide,* Essex: Pearson Education Limited.
15. Hakim, C. (1987) *Research Design: Strategies and Choices in the Design of Social Research,* London: Unwin Hyman.
16. Hussey, J. and Hussey, R. (1997) *Business Research: A practical guide for undergraduate and postgraduate students,* Hampshire: Palgrave.
17. Jick, T. J. (1979) Mixing Qualitative and Quantitative Methods: Triangulation in Action, *Administrative Science Quarterly,* Vol. 24, December, p602-11
18. Kane, K. (1985) *Doing Your Own Research: Basic Descriptive Research in the Social Sciences and Humanities,* London: Morion Boyars.
19. Kane, E. and Brun, M.O. (2001) *Doing Your Own Research,* London: Marion Boyars.
20. Kumar, R. (1996) *Research Methodology: A Step-by-Step Guide for Beginners,* London: SAGE Publications
21. Philip, B.S. (1971) *Social Research: Strategy and Tactics (2nd edn),* New York, The Macmillan Company.
22. Riley, M., Wood, R. C., Clark, M. A., Wilkie, E. and Szivas, E. (2000) *Researching and Writing Dissertations in Business and Management*, Australia: Thomson.

23. Sarantakos, S. (1998) *Social Research (2nd Edn)*, China: Palgrave.
24. Saunder, M., Lewis, P. and Thornhill,, A. (2003), *Research Methods for Business Students,* Edingburghgate: Pearson Education Limited.
25. Thietart, R. A. *et al* (2001) (translated by Samantha Wauchope) *Doing Management Research: A comprehensive Guide,* London: Sage.
26. Ticehurst, G. W. and Veal, A.J. (2000) *Business Research Methods,* Australia: Pearsons Education Pty Limited.
27. Thomas, A. B. (2004) *Research skills for Management Studies,* London: Routledge.
28. Tomlinson, J. (2006), *Experiment and Quasi-experiment* (unpublished essay), (LUBS), Leeds: University of Leeds.

Websites:
http://www.ucalgary.ca/~dmjacobs/phd/methods/
http://www.elwa.ac.uk/ElwaWeb/elwa.aspx?pageid=2087
http://www.voicewisdom.co.uk/htm/triangulation1.htm
http://www.referenceworld.com/sage/socialscience/triangulation.pdf
http//www.bowensevaluation.com/papers/SinOfOmission/main.htm